TABLE OF CONTENTS

I0503527

i. PREFACE

"It has to be personal, it should matter to YOU". This book has been made for my personal project, a project in which we explore a topic of our interest and develop a product relating to that topic. When I heard this statement, I thought back to my passion for writing. And I knew that this is just the beginning of a worthwhile journey...

Before we dive into the topic of "migration and health impacts", let me introduce myself. To begin with, I am Neha Shiju and I am 15 years old (at the time of writing this). My hobbies include painting, reading, writing, and so many more. I always try to find something new to explore. And by reading this, you are engaging with the beginning of my writing journey.

I lived in India for 2 years before moving to the Netherlands. And I am from South India, Kerala specifically. A state often called "God's own country", which is quite ironic since Kerala is not a country but in many ways it is so different that it could be a country in itself. As I travel through different parts of Kerala, I see how diverse it is. Despite speaking the same language, there are various dialects in different areas. However, these multiple religions but still share many common festivals, such as onam (a hugely celebrates festival)! After moving to the Netherlands, I attended a Dutch school for a few years and hereafter I moved to an international school. I hope this gives you some kind of idea about me. I could say so much more but the book is not about me! The book is about migration and how health is impacted by this movement. I chose those topic seeing how much the movement from India to the Netherlands impacted my family and I. We faced health problems and this always made me curious. Why was this happening? Was it actually because of the move or something else? Are others facing the same thing? What factors cause this? All of these questions motivated me to find answers and share them. This book is only a stepping stone for answers but it certainly cleared up some doubts. However do keep in mind that all of the information presented in the book is from research. DO NOT make any immediate changes or follow any advice before consulting with a professional. I look forward to you exploring this book and I hope it is enlightening.

(Eveleigh).   (Tea from Culture Tourist).

i. Preface

Since a young age, I have been interested in writing. Writing a short story, novel, non-fiction book etc. My passion for writing has only grown from there. Writing gives me a sense of freedom to express myself in the way I want to. I always write, even if it's just a reminder to do something or to plan out my holiday. It is a powerful tool that everyone can develop and use. My goal with this book was to write something original and interesting and this is my first attempt at it. Whether this book becomes famous or not, it means just as much to me. I have spent months working on this book, not only because of my passion but because it is personal. This book applies to me, it applies to my family, and hopefully some other individuals. Although the book it meant to be specific, I have broken it down in a way that makes it more relatable and useful, giving more people the opportunity to see how it connects to them.

Migration your experiences

"write more than you could ever say"

"and leave behind a legacy..."

·····················Write your Interests·······························

Use this as your personal journal!! You will be seeing a lot more of personal writing in this book!

What interests do you have? What hobbies do you engage in and why?

....
....
....
....

Consider which of these (or which 2 or 3) you would like to further develop. And how would you do this?

....
....
....
....

ii. To know before reading:

Social and political factors that cause human migration include threats, wars or humanitarian crises that people face which force them to leave their country, these are examples of push factors (European Parliament 2).

Demographic and economic factors comprise greater job and education opportunities and better living conditions.

Most people aim to move to more economically developed countries (MEDCs) from less economically developed countries (LEDCs). However people can also move from MEDCs to MEDCs to experience a different life or gain new experiences. These are pull factors as they are factors which lead someone to seek refuge in another country for greater benefits and advantages (European Parliament 2).

Lastly, environmental factors are push factors which oblige people to leave for the better. Natural disasters create unstable and risky environments. Examples of some natural disasters are tornados, hurricanes, cyclones, floods and volcanic eruptions (European Parliament 2).

Most reasons for human migration fall within one of these categories.

The types of migration are: internal migration, external migration, seasonal migration, emigration, immigration, chain migration, cyclical migration, economic migration, environmental migration, forced migration, step migration and voluntary migration (Sawe). The type of migration depends on the person's personal situation as well as the country's state and opportunities.

Internal migration means staying within the political borders but shifting the location (Hunter). For example, going from one place to another within a region, country or continent.

ii. TO KNOW BEFORE READING:

External migration means moving outside the political borders to a different country (Hunter).

Seasonal migration refers to temporary migration of people for more economic prosperity and living conditions (Sawe). This type of migration is often undertaken by farmers to support their livestock by providing it with the right temperatures and circumstances (Sawe).

Emigration is abandoning the original country of settlement and immigration is the movement to a new country (Hunter).

Chain migration is when family members migrate to the same destination ("Chain Migration"). Usually one or two members of the family establishes themself in the country and then sponsors other members ("Chain Migration").

Cyclical migration is when someone temporarily leaves their country for a short period of time and it is generally due to their job (Datta 1).

Economic migration involves the movement of people from the original location to a new place for greater economic advantages (Kyoko Kusakabe).

Environmental migration is the movement of people due to sudden natural catastrophes that can cause potential negative health impacts. They are refugees since they are fleeing involuntarily (Dun).

Forced migration is the involuntary movement of people due to unforeseen environmental, social (and political) or demographic issues. (Simeon et al.).

Step migration is temporarily settling in a place and then moving to other countries at a later stage. This is usually for economic advantages (B.).

ii. TO KNOW BEFORE READING:

Voluntary migration occurs when people choose to change their place of residence for improving certain aspects of their life ("Types of movements"). This type of migration is more reliant on pull reasons as there may be countries with traits that attract people to settle there.

Additionally, the major factors impact health are your lifestyle, insufficient nutrients and exercise, stress, culture and genetics.

Your physical state, mental state, beliefs and knowledge and inherited traits and genetics are what majorly impact your health.

Since this book covers health impacts of migration, it is also important to consider what factors affect health. However, it is important to note that mentioned effects will impact individuals in the distinct ways as our bodies are different. It is also crucial to realise that this book is written by a student, not a medical professional so all advice should be taken lightly. This is a research-based book. Before you choose to make any changes to your lifestyle, please consult a medical professional.

With that being said, please remember that while this book is written by a student, the information is **valid** and **credible**. Information presented in this book is based on research done by professionals and academic websites. All information should be easy to trace back using the MLA9 in-text citations and work cited page. Lastly, if you are confused regarding any of the terminology used, please refer to page (page number) and check the index for the term.

This should be kept in mind when reading this book.

Now that this has been addressed, thoroughly enjoy the book!

iii. INTRODUCTION

Migration is inevitable in cities where life is demanding and our stability is questioned daily. Whether that is in terms of health, financial status, safety or happiness. Human migration is when people permanently change their settlement, moving from one geographical location to another. This decision can be made for a variety of reasons and there are also different types of human migration. The reasons for human migration can be classified into social and political factors, demographic and economic factors and lastly environmental factors. These factors can either act as push or pull factors, push refers to them driving people away from their original settlement whereas pull factors give people a reason to migrate. See the previous page for a detailed explanation of types of push and pull factors.

 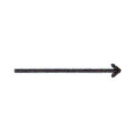 

(History.com editors).   (Kiffel-alcheh).

In this book, it will be discussed what effect migration has on our health. Why is health affected? What are implications of migration? And what can we learn from this information? All of this will be covered in the book. As a school student who is passionate about science and wanted to explore a personal topic, I chose the topic of migration as it is a common phenomenon that people go through and I have also experienced. The book covers various topics. Here's a little sneak peak into the table of contents!

TABLE OF CONTENTS

The situation being focused on in this book is the voluntary movement from a hot climate to a cold climate as this is what I am familiar with. I was always curious to know why some are affected by this change while others aren't. Is it the universe targeting random individuals or can it be explained by science after all?

The Positives:

Physical health is one aspect to consider when discussing how migration affects us, however, mental health is equally important. Poor mental health can also negatively impact physical health. Balance is the key and this balance is often lost when moving from one country to another but some positive effects also exist. Here it will be evaluated whether mental health is positively or negatively affected due to migration. This page covers the positives.

1) An exciting adventure
As Lao Tzu said "A journey of a thousand miles begins with a single step" (Nucleus AI). Moving to a new environment can be a sudden but it certainly brings an element of surprise and excitement that makes it fun and enjoyable for some ("Could Moving").

(Chowdhury).

2) New language, new culture
Learning a new language is always useful. It is a skill that will benefit you in the long run if it is practiced well. Additionally, engaging in a distinct and unique culture will give you a new perspective and allow you to be accustomed to their way of life ("Could Moving"). In life we will always have to endure new changes but it isn't always that bad right?

3) A creative approach
It is said that moving or travelling boosts creativity (Christian). A study by the American Psychological Association also supports this ("Living Outside"). Increased creativity is especially useful for professions that require new, out-of-the-box thinking.

The Positives:

4) A new social circle

When moving to a new place, you will encounter new faces and form new relationships. The MayoClinic has found that it can also ensure belonging which also boosts happiness (Rob). This is one of the benefits of a new social circle.

································ Your experiences ································

It is always important to reflect on your past experiences and take them into consideration when making decisions.

Have you travelled before and felt that it impacted your mental health positively?

....
....
....
....

If yes: How did that make you feel overall/when you travelled back?

....
....
....
....

The Negatives:

Here it will be evaluated whether mental health is positively or negatively affected due to migration. This page covers the negatives.

1) A daunting adventure

Humans are dependant on habits as it is a tool that allows us to reach our goals and make ourselves more focused (although it can also be the opposite) ("Understanding Habits"). This makes change unprecedented and daunting at times. Having to rely on the unknown is difficult for everyone due to the way we function. This is one of the downsides of movement to a new country/place.

2) Culture shock

When first moving to a different country, it is normal to experience culture shock. Discovering habits that you would never follow or things you find "uncommon" should suddenly become the new normal and adjusting to the culture may take some time and effort. Culture shock can also cause anxiety and a feeling of estrangement ("What Is Culture"). ; (Verto Education).

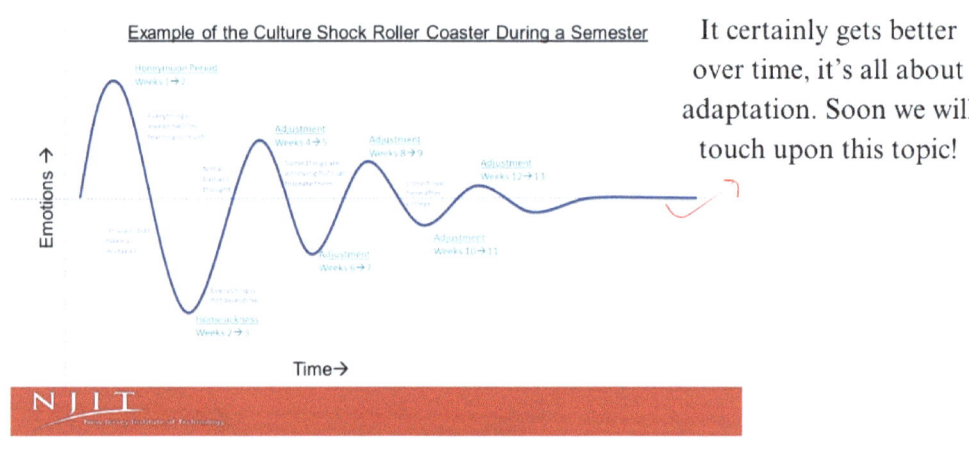

It certainly gets better over time, it's all about adaptation. Soon we will touch upon this topic!

("Cultural Shock").

3) Anxiety, stress

When moving to a new place it is also normal to go through anxiety and stress. We experience this because we are not aware of what the future holds and this causes us to contemplate what may be coming next ("DEALING WITH").

4) Leaving friends and family

One of the biggest reasons people feel emotionally exhausted is because of haivng to leave friends and family. People we have spent lots of quality time with and maybe even grew up with, will now no longer be part of our life. This change can provoke anxiety and depression, negatively impacting our mental health (Army Maj and Mary Markivich).

The Negatives:

5) The Climate

It is hard to say in a generalised manner how a change in climate can negatively affect our mental health however some examples will be mentioned. One such example is that poor air quality can have effects such as depression and anxiety (Massazza). Climate change also affect mental health negatively, from feeling anxiety to post-traumatic stress disorder (PTSD) (Hough and Counts). This is due to the adverse effects of climate change on daily life such as floods, hurricanes, drought etc.

Is sunlight associated with happiness? Consider what hormone is released upon exposure to sunlight.

If sunlight is associated with happiness, then what is the cold, winter associated with? And is it valid and reliable to draw such a conclusion?

(Jordan).

······························· Your experiences ·······························

It is always important to reflect on your past experiences and take them into consideration when making decisions.

Have you travelled before and felt that it impacted your mental health negatively?

....
....
....

If yes: How did that make you feel overall/when you travelled back?

....
....
....
....

DIET AND ITS ROLE

The Positives:

Diet is crucial to maintain a healthy and balances lifestyle. A healthy diet can keep all our organs safe and healthy whereas a poor diet can weaken our body making recovery difficult and painstaking. So here it will be evaluated whether diet is positively or negatively affected due to migration. This page covers the positives.

1) Food diversity

Every country has its own cuisine. By moving to a new country we are given the option to try new variaties of food and experience the richness in culture. This may also be advantageous for our health as food variety is useful to absorb nutrients that were previously not absorbed (Relocation).

Look for a rainbow of colours in your food! That's when you know you are getting all the essentials.

(Davoli).

2) A balanced diet

Although exploring new food does not mean that we will automatically become healthier, we can control how well balanced our diet is. Furthermore, moving to a new country does allow us to make new lifestyle choices rather than sticking to habits that we know are unhealthy. A balanced and healthy diet can even do wonders such as improving our focus and helping with depression and anxiety ("Eating Well"). ; ("Food for your").

3) New shopping habits

As mentioned earlier, movement to a new country allows us to make new lifestyle choices. Shopping habits are crucial to manage because they directly impact our diet and health ("How to have"). Purchasing food that will give us required nutrients, allow a balanced diet as well as something that satisfies our needs is what should be focused on. This can benefit our health to a great extent.

The Positives:

4) Adjusting to the staple food

Adjusting to the food of your country is not easy as it may be very different from the food of your country of origin/country you lived at before. For example, if you come from a place where food is always fresh and you move to a new place and only eat processed food it can drastically impact your health (Holmboe-Ottesen and Wandel 34). When making a big change to our diet, it should be a slow transition and we should try to incorporate some previous elements of our life so that the change is not too sudden.

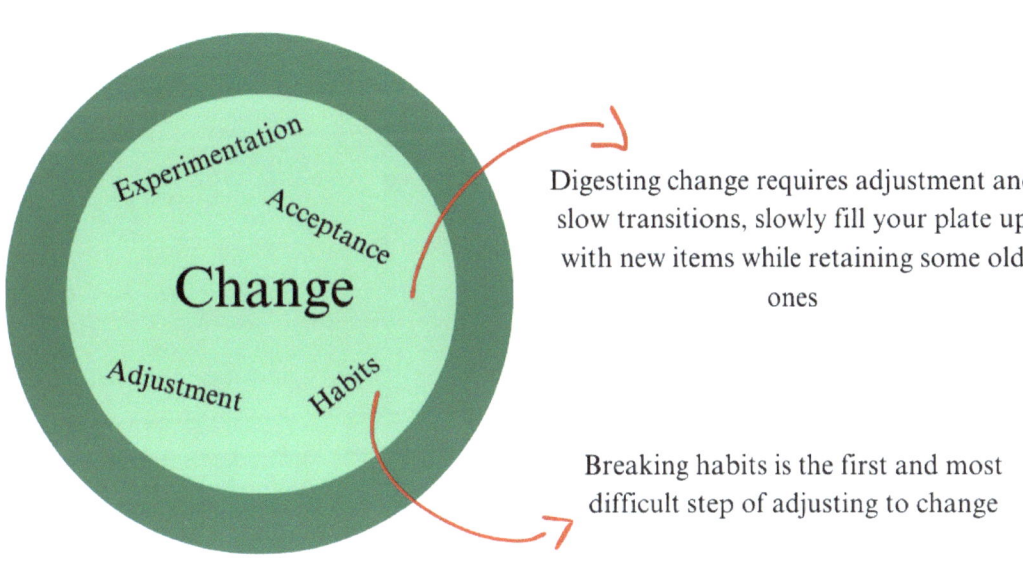

Digesting change requires adjustment and slow transitions, slowly fill your plate up with new items while retaining some old ones

Breaking habits is the first and most difficult step of adjusting to change

·············· Your experiences ·····················

It is always important to reflect on your past experiences and take them into consideration when making decisions.

What healthy food items do you consume the most and what nutrients are you getting from them?

....
....
....
....

What can be seen as positive about your diet?

....
....
....
....

The Negatives:

Here it will be evaluated whether diet is positively or negatively affected due to migration. This page covers the negatives.

1) Exotic and foreign ingredients

Familiarity with local ingredients is extremely important for health. Using this information, it can be noted what one may be allergic to, what works for their body and likes/dislikes. A study by Mintel found that "27 percent of American diners prefer to order healthy meals with familiar ingredients" ("Familiar ingredients"). This goes to show that people prefer knowing ingredients well because it helps them work out what they can eat. Consequently, this is a challenge in a new location and it can be exhausting to become familiar with foreign ingredients (Relocation).

2) Change in mindset

Different countries have different values, ideas, cultures etc. An example of an idea that some cultures follow are finishing everything on your plate, this can cause weight gain but the opposite can also happen if people are encouraged to be skinny/look thin ("Impact of culture").

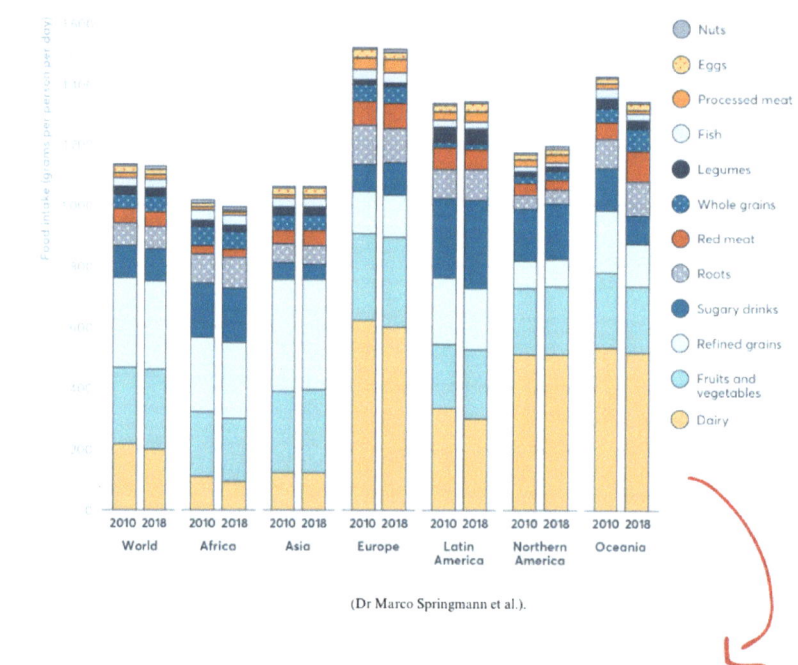

(Dr Marco Springmann et al.).

Different types of food are eaten in different proportions as the country and people vary. This results in some nutrients and minerals only being available to certain groups of people as they consume that specific ingredient.

The Negatives:

3) Gut bacteria

There was a study published in the journal Cell and it found that bacteria in our body can be affected when moving to certain countries. The bacteria in our gut is more vulnerable to problems, for example gut microbiome diversity decreases due to migration and this can give rise to different health issues (Tousignant). ; (Chloe James and Ian Goodhead). There is also more chance that those who migrate become obese rather than native individuals and this is because of the underlying complexity of their body and the way they grew up (Chloe James and Ian Goodhead).

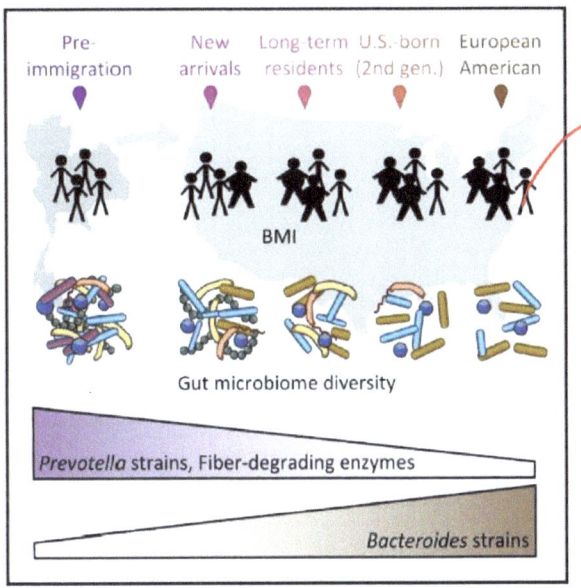

As you migrate, your gut microbiome diversity decreases, causing various other side-effects and harming the body

(Chloe James and Ian Goodhead).

···································· Your experiences ····································

It is always important to reflect on your past experiences and take them into consideration when making decisions.

Are you eating too much unhealthy food in your day-to-day life?

....
....
....
....

What can be seen as negative about your diet?

....
....
....
....

3 THE POWER OF CLIMATE

Climate plays a crucial role in our health. It is one of the determining factors of whether our body can adjust to a new environment. However unlike the previous topics, it is harder to evaluate because of its complexity and the circumstances. Taking this into consideration, the hot to cold climate movement scenario is provided (as this is what my family experienced and this is more common) and it is discussed based on the circumstance.

Hot to cold climate movement

When moving from a hot to cold climate within a short period of time there are multiple difficulties that our body encounters. One such difficulty is organs drying out, such as eyes and skin (Tiwari). The reason for this is because the ambient temperature indoors is not at the same level as the temperature outside (Tiwari). In order to adjust the temperature inside, we use artificial heating methods that warm up our surrounding air but this is poses such problems. A method of heating that has been known to be dangerous is fuel burning heaters ("Why Indoor"). They should be used mindfully as they emit carbon monoxide ("Why Indoor").

Some other consequences of this type of migration are colds, asthma, muscular pain etc. (Tiwari). The Allergy and Asthma Network also found that a difference in temperature (going from hot to cold) can worsen asthma and trigger flares ("Cold Air and Asthma"). It is also crucial to mention that those with existing underlying diseases should be even more cautious as some of the diseases can worsen. Asthma is one example but arthritis can also flare up due to potential joint pain (which is more common in the cold) (Greenfield). ; ("Arthritis Flare-Ups").

It goes without saying that the sun is not present as often during the winter. This can challenge our physical wellbeing and mental wellbeing. The decrease of sunlight can affect our energy, diet and concentration ("Pros and Cons").This is called as seasonal affective disorder (SAD) or "winter depression" ("Pros and Cons"). Our happiness also tends to decrease during winter times as our serotonin declines ("Pros and Cons"). These are some general implications on your health when moving to a hot climate to a cold climate.

·· Your experiences ···

It is always important to reflect on your past experiences and take them into consideration when making decisions.

What is the climate like in your region and how well do you adjust to it?

....
....
....
....

In what ways has climate affected your body in the past?

....
....
....
....

4 NATURE'S GIFT

Our body and its capabilities are invaluable and we should always be thankful for that. By virtue of evolution, we can adapt better than before which allows us to travel all over the world. It also allows us to adapt in a new place in the long term, otherwise we would not be able to settle in a place other than our birth place. The University of Michigan found that Homo sapiens can thrive in all habitable biomes which is a huge advantage for us (Sherburne). An example of adaptation to survive in distinct environments are how the Tibetans are able to live in high altitudes, at which the oxygen level is 40% less than sea level (National Geographic Society). The way this works is an increase in hemoglobin so that more oxygen is distributed around body (National Geographic Society).

(Lehnardt).

Humans are referred to as "homo sapiens", but what does this actually denote? Homo sapiens refers to being a "wise man" (Tattersall). The scientific name of modern humans is homo sapiens sapiens (Dorey). There is a repeat of sapiens here as there are two different species of homo sapiens. One is the homo sapiens sapiens (modern humans) and the other is homo sapiens neanderthals (Stringer).

It is through the process of evolution that we developed the required characteristics to survive on earth. Evolution is based on natural selection in which favourable traits are passed on through generations and the "fittest" survives and continues reproducing ("What is evolution?").

Charles Darwin (a naturalist) was the one that proposed theories about evolution and natural selection which gave us insight into how these processes work (Desmond). There are 5 important concepts (VISTA) that Darwin explained regarding natural selection, these are as follows:

VISTA
- Variation and Inheritance

Every human is different, so is every animal. In the process of natural selection, change occurs in the DNA (deoxyribonucleic acid) ("How Does"). These slight variations are called mutations, they are not always harmful but they can be in some cases (Loewe). An example of a harmful mutation is when a disease is passed on through a mutation, an example of a beneficial mutation is camouflage for animals.

This is the study of taxonomy, extremely interesting!

Classification of *Homo sapiens* within the order Primates

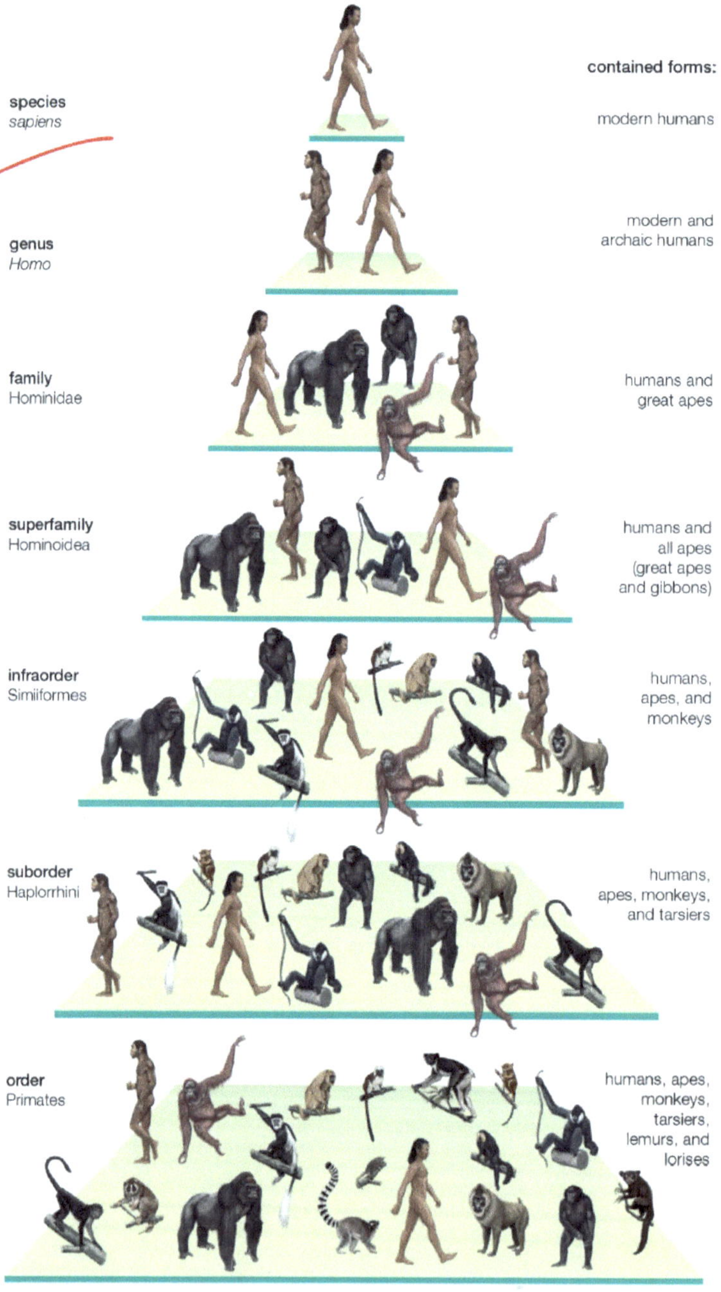

Humans are "homo sapiens" because their genus is homo and species is sapiens. The genus and speicies forms the scientific name for species.

		contained forms:
species *sapiens*		modern humans
genus *Homo*		modern and archaic humans
family Hominidae		humans and great apes
superfamily Hominoidea		humans and all apes (great apes and gibbons)
infraorder Simiiformes		humans, apes, and monkeys
suborder Haplorrhini		humans, apes, monkeys, and tarsiers
order Primates		humans, apes, monkeys, tarsiers, lemurs, and lorises

("Homo sapiens").

Inheritance refers to the passing on of genetic material to the offspring. This genetic material determines the traits of the offspring, such as eye colour, height, skin colour etc ("How Does").

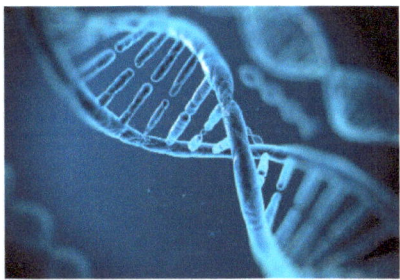

("8.2% of our DNA is 'functional'").

- Selection

Selection is used to describe "survival of the fittest". Only those who thrive in their given habitats can reproduce and continue their existence. Characteristics that favour ones habitat and protect one from predators will have an advantage in this survival. Species that are not able to survive will go extinct over time. It is surprising to find out that 1,000,000 species could potentially go extinct ("UN Report"). An example of an extinct species is the woolly mammoth (Hilfrank). It went extinct due to rapid heating of our planet which it could not cope up with (Mathews). This shows how animals only survive if they can adapt with their surroundings.

Types of natural selection:

Disruptive selection

Disruptive selection is the concept that explains how extremely different and unique species breed more than regular ones, this usually occurs in places were there is more competition for survival (Martin and Pfennig). An example of disruptive selection is the peppered moth, the two extreme variants of the peppered moth will reproduce more than regular moths in this case ("Disruptive Selection").

("Disruptive Selection").

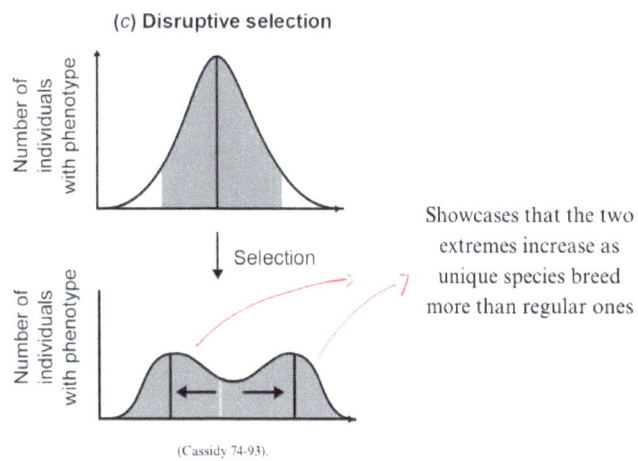

(c) Disruptive selection

Number of individuals with phenotype

↓ Selection

Number of individuals with phenotype

Showcases that the two extremes increase as unique species breed more than regular ones

(Cassidy 74-93).

Directional selection

The second type of natural selection is directional selection. This is when a species with one specific trait has an advantage with regards to their survival and consequently, is preferred more than the other species (MacColl). You could think of this as leaning towards only one direction, only one species that is better fit for survival is preferred. An example of directional selection is giraffes with long necks ("Natural Selection"). Giraffes with long necks can survive better as they are able to obtain the food they require whereas giraffes with short necks are not as easily able to do that. This allows the long-necked giraffes to survive in their environment.

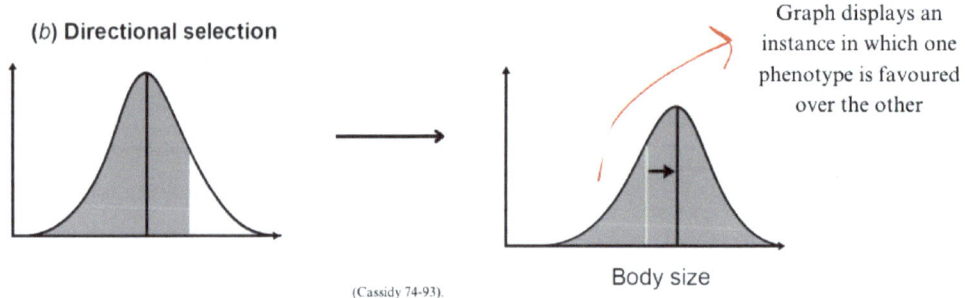

(Cassidy 74-93).

Stabilising selection

The last type of natural selection is stabilising selection. This can be thought of as the opposite of disruptive selection as average or common species are preferred more than extreme species. An example is the average height of plants, they need to have an average height for the required sunlight so this type of selection helps with that ("Natural Selection").

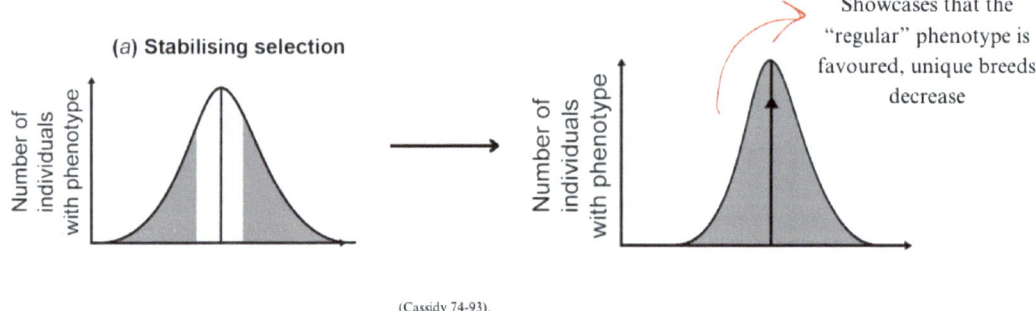

(Cassidy 74-93).

Time and Adaptation

Over time species are able to develop characteristics that make their survival easier in a particular location ("How Does"). Natural selection helps determine which species will continue their survival and what genes will be passed on. These 5 points form the concept of natural selection that allows individuals to thrive in their environments. This is why humans are able to survive in hot or cold places without dying. Although it is true that we may face underlying health issues, we are certainly more adapted than species 1000 years ago. This is what makes homo sapiens special and unique. We are a product of multiple centuries of evolution, natural selection and adaptation. We should be thankful to nature's gift.

Around 710 million people would migrate if given the chance (Esipova et al.). Whether this is because of finding and experiences diverse cultures, for financial advantages or for familial responsibilities is unknown but this value shows many people find it hard to settle in one place for an extended period of time. There are a few more concepts that are useful to know when discussing migration, these are:

Brain Gain

Brain gain is when experienced workers move for financial advantages or opportunities ("brain gain"). However when thinking critically about this situation, one may consider whether this is fair. Skilled individuals are abandoning their home country and go to work in distant countries. Well, if these individuals return to their original country, it is called "brain circulation" ("Brain drain"). It forms a circle as you go from your original country to another country and then return back to your original country.

Brain Drain

Brain drain is the opposite of brain gain, skilled workers emigrate leaving behind a country (Young). This can affect a country's economy negatively as these skilled people's work is being taken to another country. Brain drain has even negatively impacted the EU as it has lost valuable people to the concept of migration (Ebbesen).

Potential Net Migration Index

The potential net migration index (PNMI) gives an estimated percentage value that shows the population brain drain and brain gain of a country ("Potential Net Migration"). This index takes into account how many people would like to move to a new country if given the opportunity, how many people would like to move into the country and taking the adult population into account (Esipova et al. 4). If a country has a high PNMI value, it signifies that more people would like to move into the country. The graph below should the PNMI values for different countries ("Potential Net Migration").

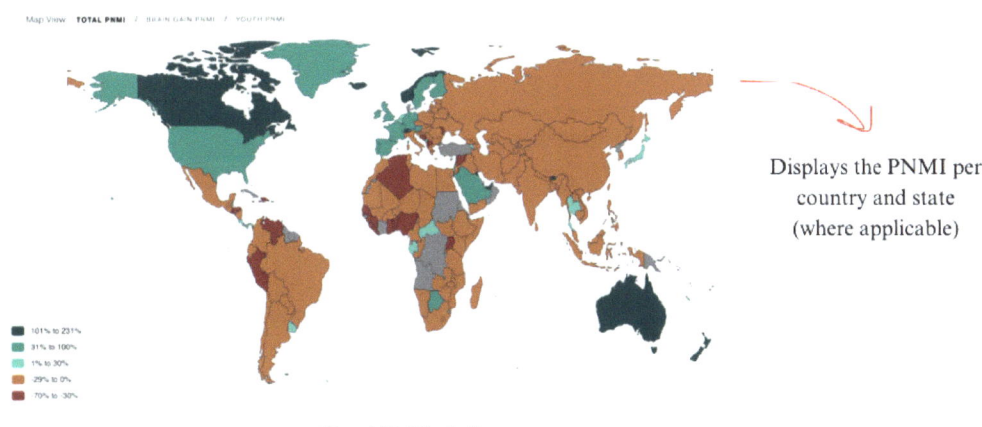

Displays the PNMI per country and state (where applicable)

("Potential Net Migration").

Stories of moving to another country:

There are 4 case studies/stories provided below that different people shared and it covers different and diverse experiences around the concept of migration. These are real experiences which can be helpful to get a glimpse of their lives and how movement has impacted them positively or negatively. Please note that these are anonymous so respect the privacy of those who have shared these experiences, this information is used to look at some real-life experiences to get a glimpse of migration and its effects:

1: Cold-hot movement and health

1. "My health has improved by leaps and bounds. The treatment of asthma and COPD is actually aimed at combating symptoms, rather than curing them. So moving to a tropical, more or less dry environment has done me a lot of good. I can't recommend it to everyone, as it would become too busy at tourist points around the world. But still, only positive!" (Shiju).

2: Remigrating and health

2. "Well, if you as a Dutch citizen remigrate to the Netherlands, just like emigrating to the continent of our all origin Africa, you will suffer for some time from a change in the viscosity of your blood, which quickly adapts to climatic differences. And this also applies, for example, to sleep adjustments due to changes in day-night rhythms. Or, for example, susceptibility to malaria." (Shiju).

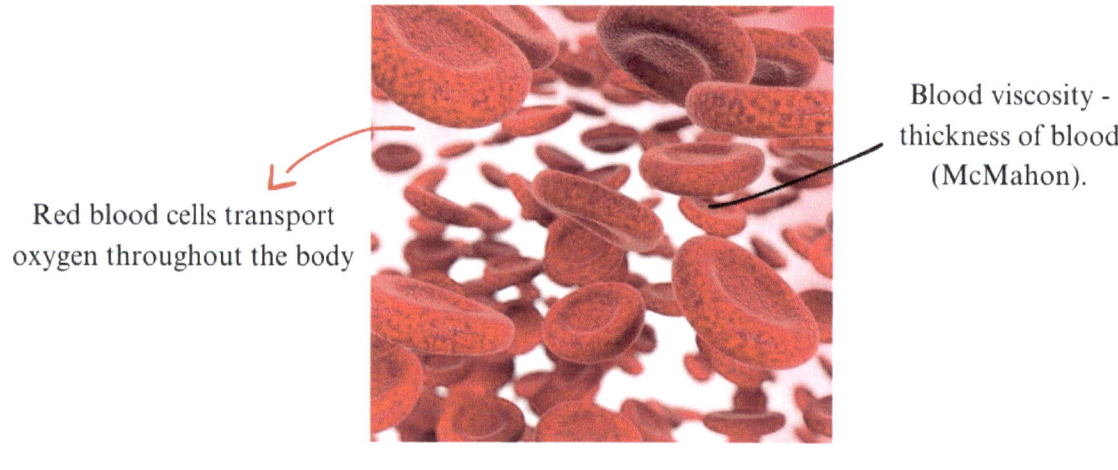

Red blood cells transport oxygen throughout the body

Blood viscosity - thickness of blood (McMahon).

(McMahon).

3: Migration and allergies

3. "Migrating from one climate to another can affect your health if you suffer from allergies to certain plants, trees/hay fever, etc. Since my childhood, I have lived alternately in countries with a tropical savannah climate, tropical rainforest climate and the well-known temperate North Sea climate of the Netherlands. In addition, short, but especially longer holidays (up to 4 months) in countries with a Mediterranean climate. (continued on the next page)

In the first months after moving, the complaints I had in my previous place of residence disappeared immediately as soon as I stepped off the plane. Somewhere between 6 months and 1 year afterwards: the same complaints but from other plants, trees, etc. I have undergone several allergy tests. In 2012, an Indonesian allergist in Amsterdam gave me an injection treatment. Unfortunately, I don't remember the name of the medication, but it had to be imported from Germany, with special permission from the Ministry of Health (and at my own expense). I can now live and go on holiday anywhere I want. I have been living in the Netherlands again since 2021 because my children and 3 of my 4 grandchildren live here." (Shiju).

4: Altitudes and health

4. "Coming from sea level (Netherlands) to mountain level (Colombia) (2500 meters altitude - minus 8 meters below sea level) Could climb the stairs in Colombia with great difficulty. My heart was racing. Came up panting. As if I was suddenly very old. After a few weeks I got used to the height difference. Body weight had now gone from 120 kg to 79 kg. Other food, more exercise. That with my 1.84 cm height. From flatland to mountainous land. Back in the Netherlands I went from 95 kg to 130 kg in no time. Probably also due to my sedentary profession. I can say that I felt better in Colombia than in the Netherlands in terms of health. I lived in Colombia for over 16 years. (from the age of 27) until the age of 43. I have been back in the Netherlands since 2006. I am now 60 years old. Everything deteriorates, hearing, visual field, walking and condition. I have become diabetic, polyneuropathic, etc. I recently got dentures. I am now 92 kg, my height has dropped to 1.82 cm. Would like to spend my old age in Colombia. Whether this option is available depends on a number of factors. Especially health and finances." (Shiju).

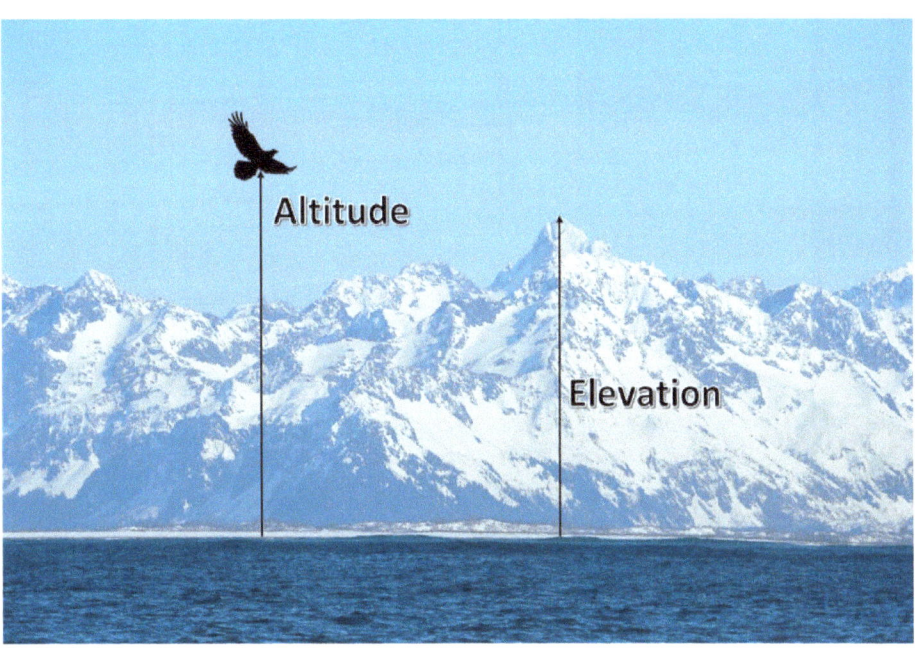

(Dempsey).

(6) Epilogue

This book has covered 5 main chapters. These chapters aimed to cover a broad range of topics that discussed migration through one or multiple lenses. In the migration and mental health chapter, there are 4 positives that were identified and 5 negatives. This suggests that overall, migration can negatively affect our mental health.

In the second chapter, there are 4 positives identified and 3 negatives identified. This indicates that in general, diet is affect positively by migration.

The third chapter is slightly different, as it only discussed one scenario. While this is primarily because hot to cold climate movement relates to me personally, it is also due to the fact that cold to hot climate has limited information. The research conducted about this scenario depicts that moving from a hot to cold climate is generally worse for our health.

The fourth chapter employs a more optimistic standpoint. It considers how we are blessed to be able to have qualities that allow us to survive in different areas, where the circumstances (such as temperature) may be vastly different.

Last but not least, the fifth chapter provides some real life experiences of people who migrated as well as covering some important information about migration (for instance, the potential net migration index). Two of the stories are promising while the other two discuss undesirable consequences of migration. Although this chapter is not particularly looking at migration as positive or negative, it covers aspects such as the impact of migration on a country's economy.

Considering all the information presented in this book, it can be understood that migration is very difficult to assess. The question of whether you should migrate may still linger on your mind, and this is completely reasonable! Migration is personal and should be determined based on what works best for you personally. So, look back over your answers for the "Your experiences" section and you may find some hidden gems. This book is not aiming to solve a problem, rather aiming to give you a sense of direction. I can't stress enough how important it is to consider what works best for YOUR body. If there is one think we need to remember when making a life-changing decision is that we are all different and we are a product of selection and diversity.

I hope this book sincerely helped you (perhaps even inspired you to engage with your passions!). I would have loved to deep-dive into some more topics but this book was written over a short period of time. I am very grateful if you have taken the time to read all of this! I wish you all the luck you deserve for the rest of your journey.

Sincerely,
Neha Shiju

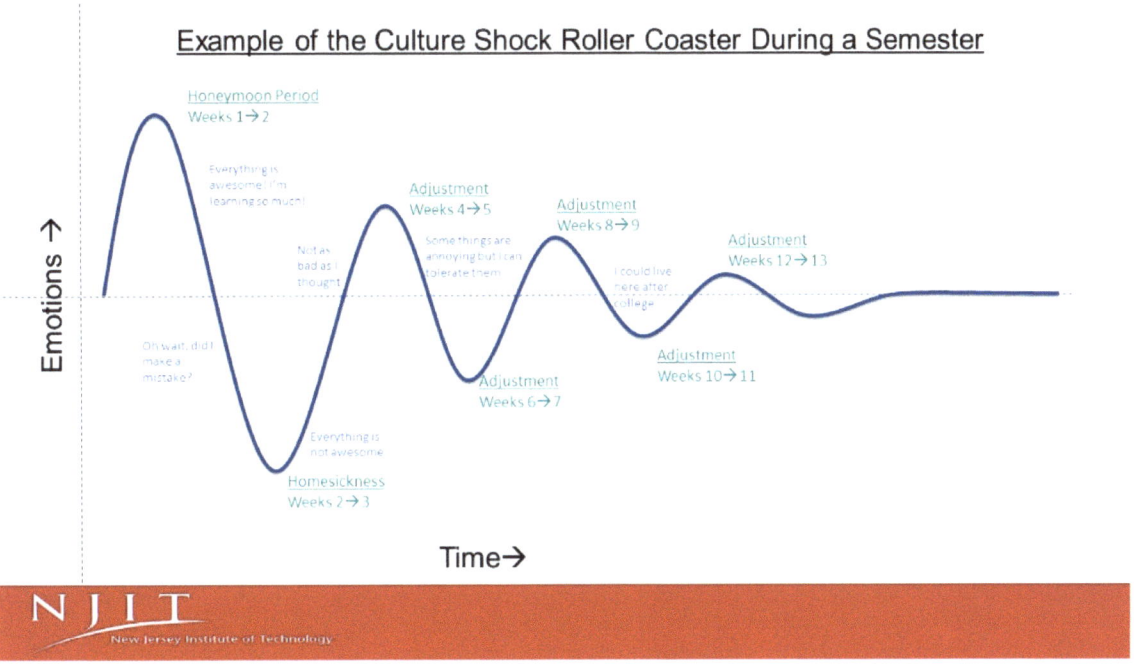

("Cultural Shock").

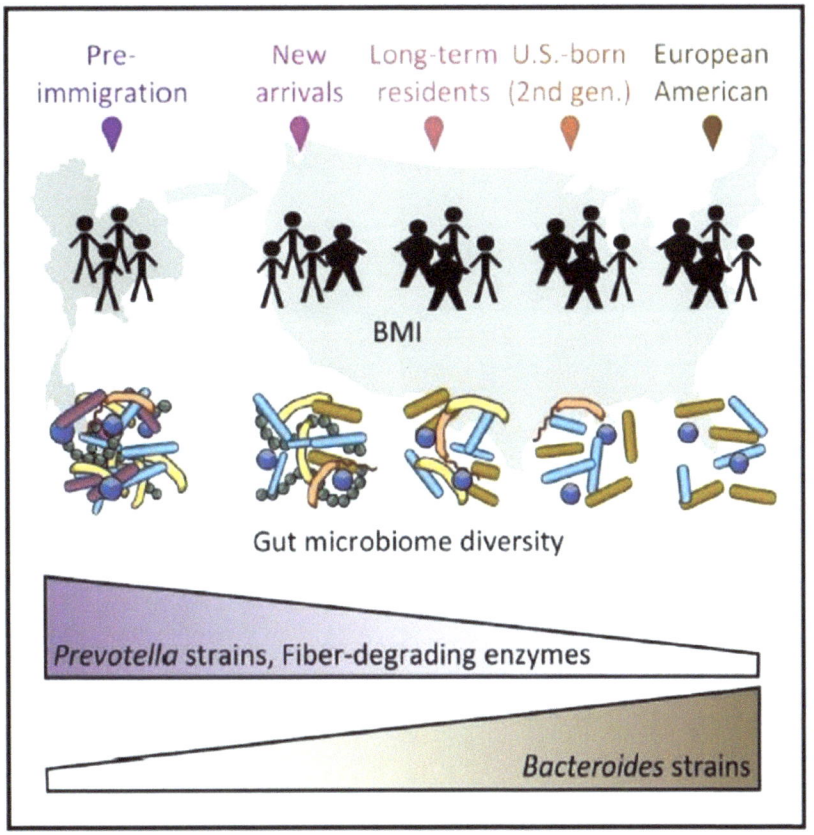

(Chloe James and Ian Goodhead).

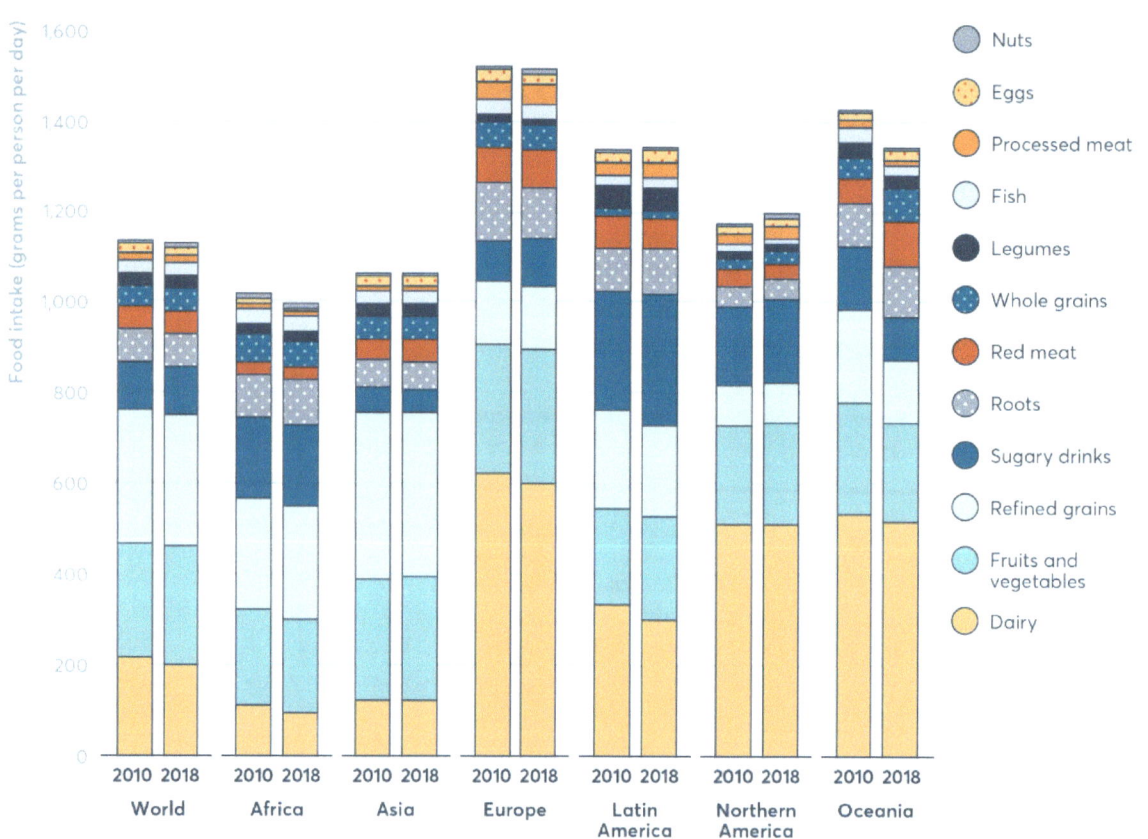

(Dr Marco Springmann et al.).

Map View: **TOTAL PNMI** / BRAIN GAIN PNMI / YOUTH PNMI

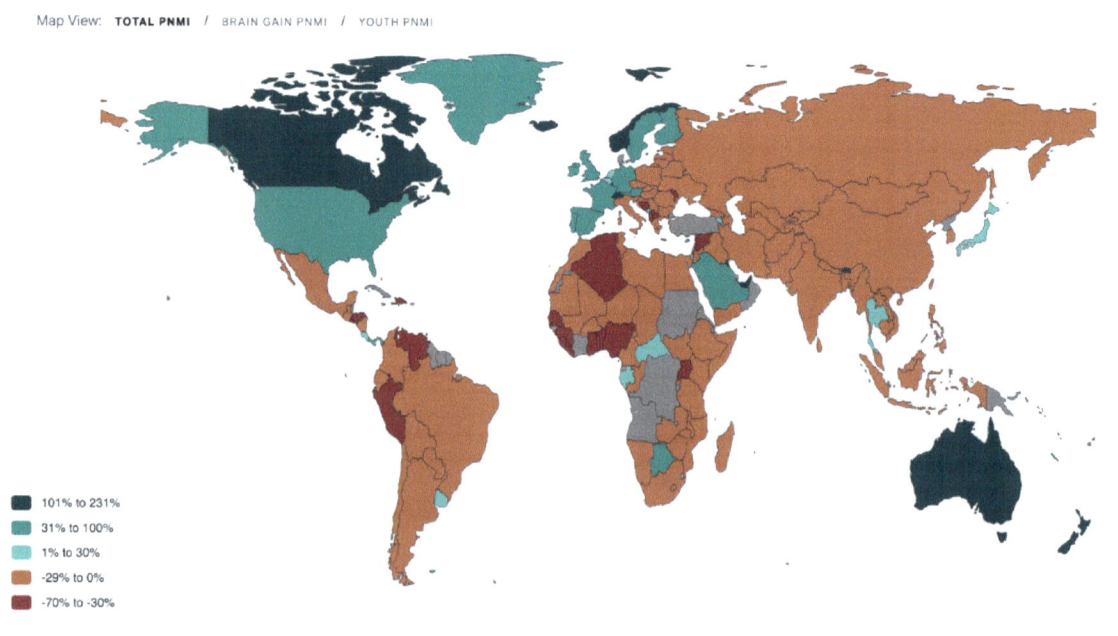

101% to 231%
31% to 100%
1% to 30%
-29% to 0%
-70% to -30%

("Potential Net Migration").

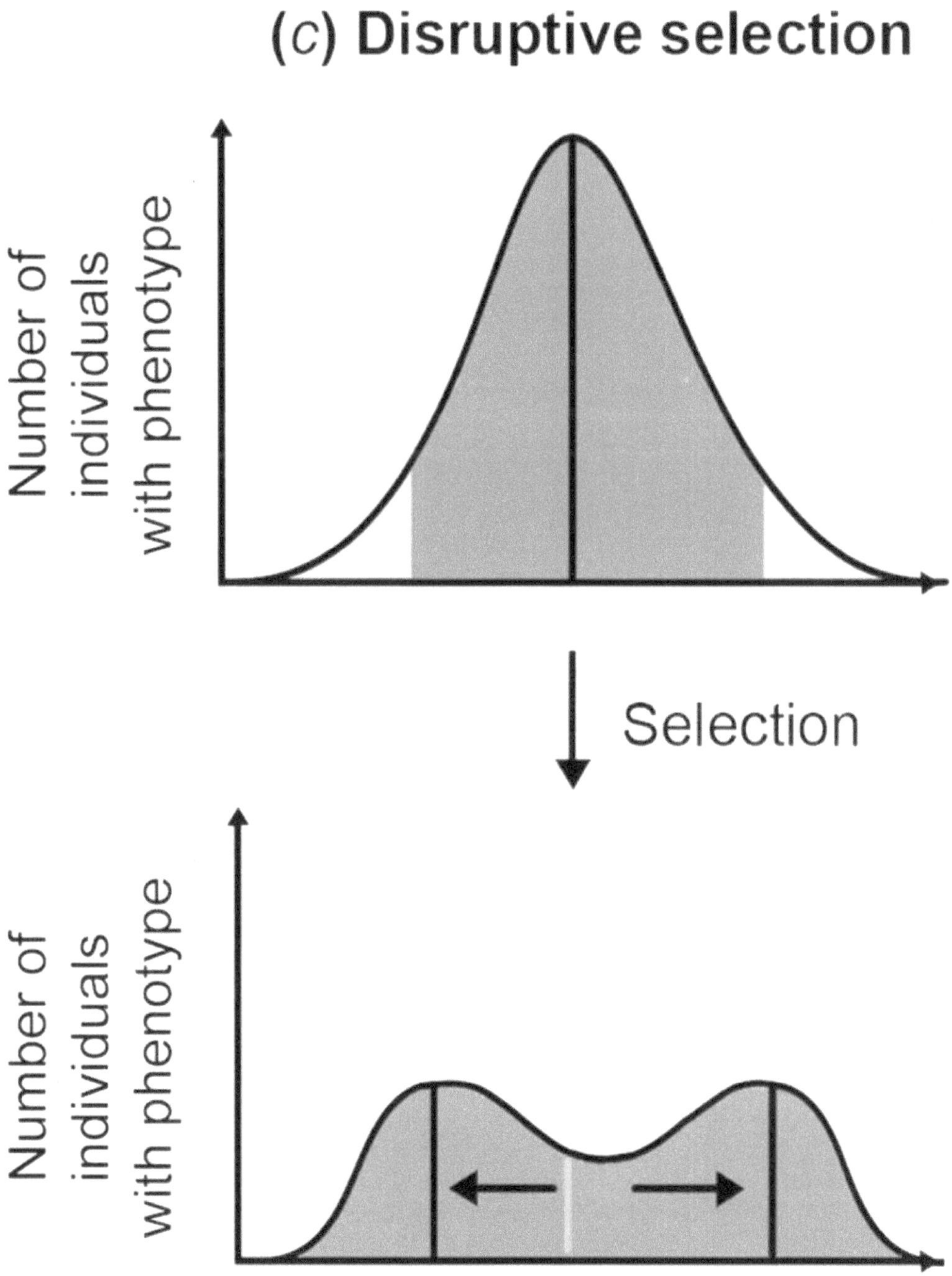

(c) Disruptive selection

Number of individuals with phenotype

Selection

Number of individuals with phenotype

(Cassidy 74-93).

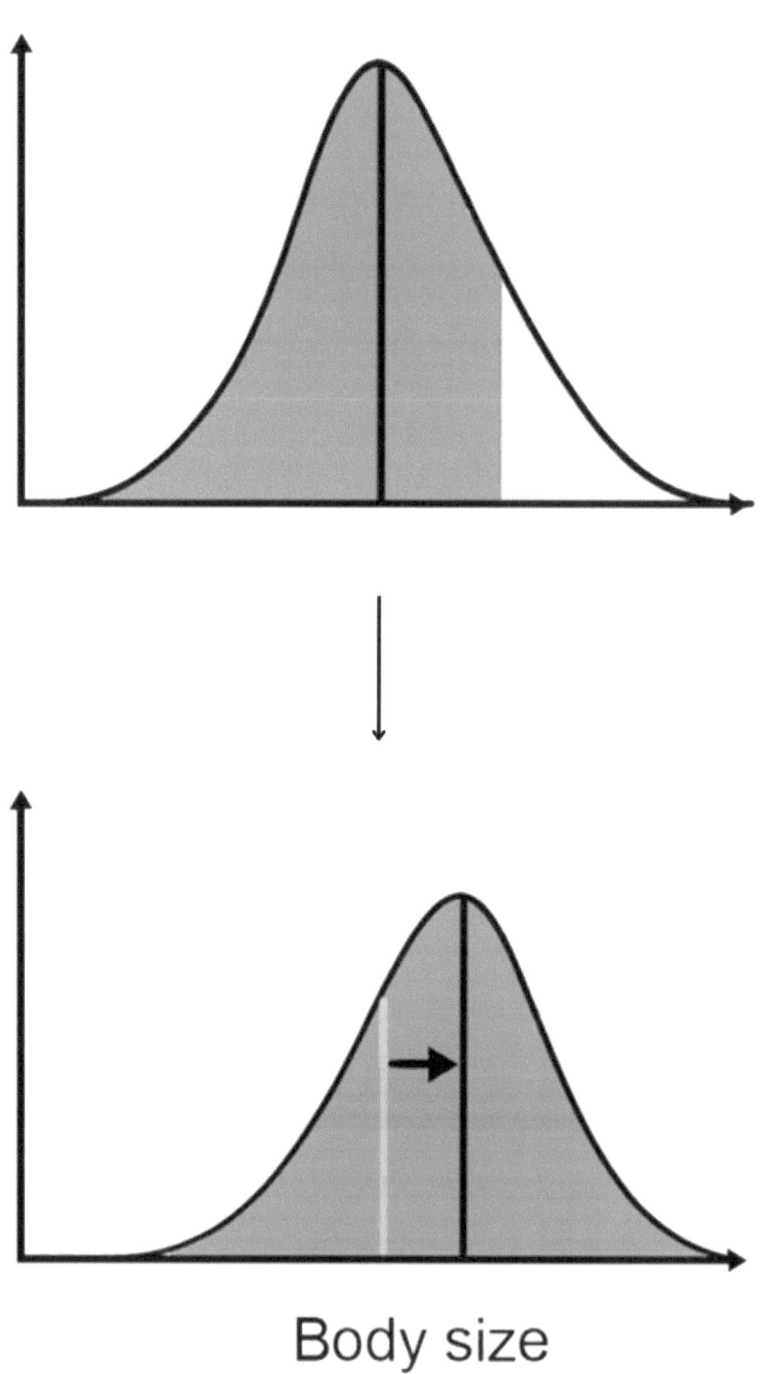

(b) Directional selection

Body size

(Cassidy 74-93).

(a) Stabilising selection

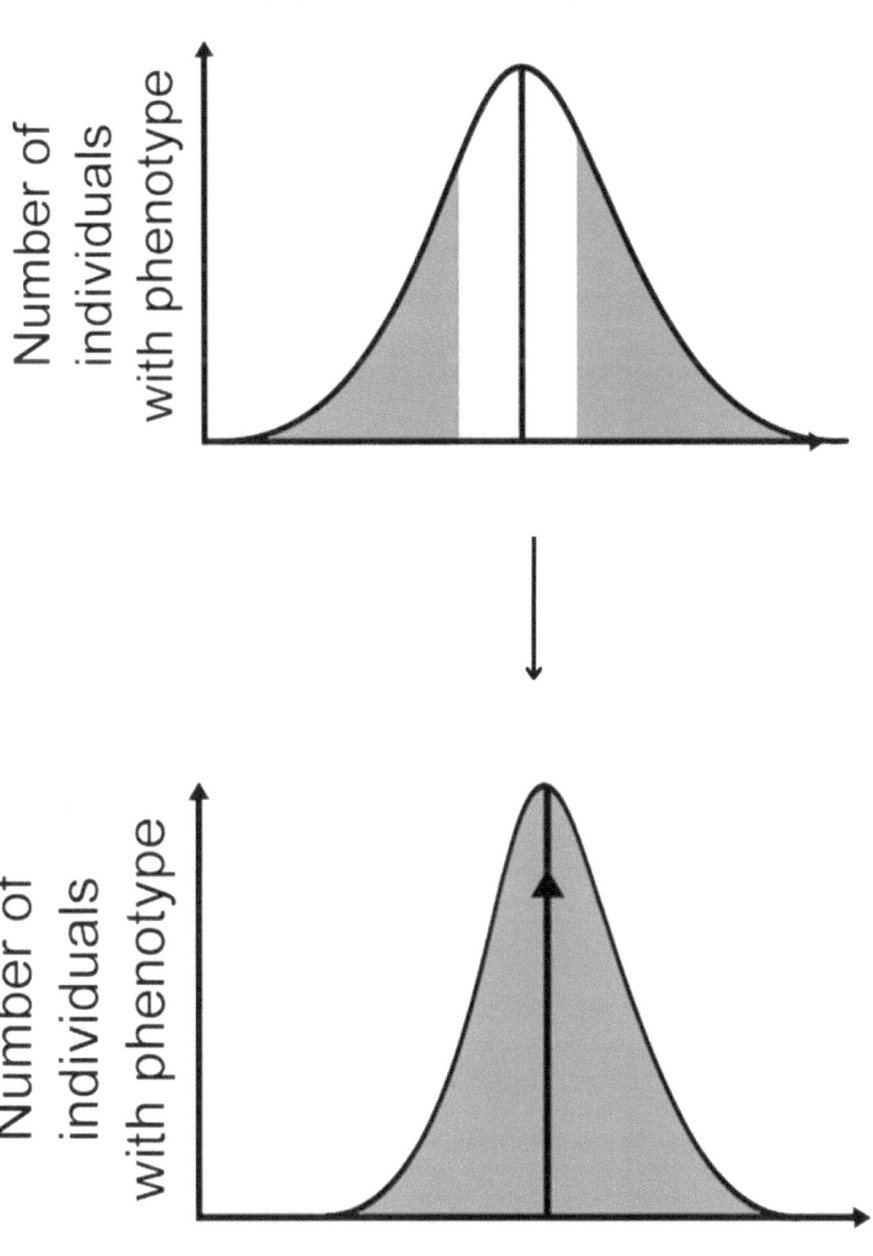

(Cassidy 74-93).

WORK CITED

Army Maj, and Mary Markivich. "Coping with Separation from Family and Friends during the COVID-19 Pandemic." Health.mil, 11 Aug. 2020, health.mil/Military-Health-Topics/Centers-of-Excellence/Psychological-Health-Center-of-Excellence/Clinicians-Corner-Blog. Accessed 13 Sept. 2023.

"Arthritis Flare-Ups: What Causes Them, What are the Symptoms & How to Find Relief." Tylenol, tylenol.com/symptoms/arthritis-pain-keep-moving/flare-ups. Accessed 14 Sept. 2023.

"Asylum and migration." The UN Refugee Agency, unhcr.org/what-we-do/protect-human-rights/asylum-and-migration. Accessed 19 Sept. 2023.

"Brain drain and brain gain." Migration Research Hub, migrationresearch.com/taxonomies/topics-migration-consequences-for-migrants-sending-and-receiving-countries-socio-economic-consequences-brain-drain-and-brain-gain#. Accessed 20 Sept. 2023.

"Brain gain." Collins Dictionary, collinsdictionary.com/dictionary/english/brain-gain. Accessed 19 Sept. 2023.

Canva. canva.com/. Accessed 20 Sept. 2023.

Cassidy, Mike. Natural Selection and Adaptive Change. Cambridge UP, 2020. Biological Evolution. Cambridge University Press, https://doi.org/10.1017/9781139016018.005. Accessed 1 July 2024.

"Chain Migration." Science Direct, sciencedirect.com/topics/social-sciences/chain-migration#. Accessed 9 Sept. 2023.

Chloe James, and Ian Goodhead. "This is how moving to another country can affect your gut bacteria." We Forum, 6 Nov. 2018, weforum.org/agenda/2018/11/moving-to-another-country-could-mess-with-your-gut-bacteria/. Accessed 13 Sept. 2023.

Chowdhury, Arka Roy. "Places in India that are perfect for adventure sports in August." Times Travel, 11 Aug. 2016, timesofindia.indiatimes.com/travel/india/travel-guide/places-in-india-that-are-perfect-for-adventure-sports-in-august/gs53497909.cms. Accessed 10 Sept. 2023.

Christian, Janet. "Living in another country is good for your brain." Medium, 19 Dec. 2021, medium.com/@JanetCh/living-in-another-country-is-good-for-your-brain-250eccd9a7f. Accessed 10 Sept. 2023.

"Cold Air and Asthma = Winter Asthma." Allergy & Asthma Network, allergyasthmanetwork.org/news/cold-air-asthma-in-winter. Accessed 14 Sept. 2023.

"Could Moving Abroad Be The Key To Improving Your Health?" Expat Focus, expatfocus.com/health/could-moving-abroad-be-the-key-to-improving-your-health-3918/. Accessed 10 Sept. 2023.

"Cultural Shock." New Jersey Institute of Technology, njit.edu/studyabroad/cultural-shock. Accessed 13 Sept. 2023.

Datta, Amrita. "Circular Migration and Precarity: Perspectives from Rural Bihar." The Indian Journal of Labour Economics, vol. 63, no. 4, 12 Nov. 2020, pp. 1143-63, https://doi.org/10.1007/s41027-020-00290-x. Accessed 9 Sept. 2023.

Davoli, Dawn. "Add Diversity To Your Meals With These 5 Globally Inspired Foods." Excela Health, 4 Mar. 2022, excelahealth.org/newsroom/2022/march/add-diversity-to-your-meals-with-these-5-globall/. Accessed 13 Sept. 2023.

"Dealing with the Stress of International Relocation." ACS, acs-ami.com/en/travel-articles/dealing-with-the-stress-of-international-relocation/. Accessed 13 Sept. 2023.

Dempsey, Caitlin. "What is the Difference Between Elevation and Altitude?" Geography Realm, 28 Mar. 2020, geographyrealm.com/what-is-the-difference-between-elevation-and-altitude/. Accessed 20 Sept. 2023.

Desmond, Adrian J. "Darwin, Charles." Britannica, 17 Aug. 2023, britannica.com/biography/Charles-Darwin/. Accessed 19 Sept. 2023.

Dorey, Fran. "Homo sapiens modern humans." Australian Museum, 16 Oct. 2020, australian.museum/learn/science/human-evolution/homo-sapiens-modern-humans/. Accessed 18 Sept. 2023.

Springmann, Marco, Mozaffarian, Dariush, Rosenzweig, Cynthia, Micha, Renata. "What we eat matters: Health and environmental impacts of diets worldwide." Global Nutrition Report, globalnutritionreport.org/reports/2021-global-nutrition-report/health-and-environmental-impacts-of-diets-worldwide/. Accessed 13 Sept. 2023.

WORK CITED

"Eating Well for Mental Health." Sutter Health, sutterhealth.org/health/nutrition/eating-well-for-mental-health#. Accessed 13 Sept. 2023.

Ebbesen, Anne Katrine. "Introducing PREMIUM_EU: A new project to prevent brain drain in Europe." Nordregio, 13 Apr. 2023, nordregio.org/introducing-premium_eu-a-new-project-to-prevent-brain-drain-in-europe/. Accessed 20 Sept. 2023.

"8.2% of our DNA is 'functional.'" University of Oxford, 25 July 2014, ox.ac.uk/news/2014-07-25-82-our-dna-%E2%80%98functional%E2%80%99/. Accessed 19 Sept. 2023.

Esipova, Neli, Pugliese, Anita, Ray, Julie. "More Than 750 Million Worldwide Would Migrate If They Could." Gallup, 10 Dec. 2018, news.gallup.com/poll/245255/750-million-worldwide-migrate.aspx. Accessed 19 Sept. 2023.

---. "The World's Potential Migrants.

European Parliament. "Exploring migration causes: why people migrate.".

Eveleigh, Mark. "Backwater cruises and ancient cures in Kerala, India's southern, sun-drenched state." Independant, 15 Jan. 2016, independent.co.uk/travel/asia/india-s-southern-sundrenched-state-of-kerala-uncovered-cruises-elephants-and-ancient-cures-a6814086.html. Accessed 21 Sept. 2023.

"Familiar ingredients, good descriptions help Americans make healthy food choices." Pizza Marketplace, 28 Aug. 2014, pizzamarketplace.com/news/familiar-ingredients-good-descriptions-help-americans-make-healthy-food-choices/. Accessed 13 Sept. 2023.

"Food for your mood: How what you eat affects your mental health." Aetna, aetna.com/health-guide/food-affects-mental-health.html#. Accessed 13 Sept. 2023.

Greenfield, Ethan. "What to know when moving to a cold climate." Moving Tips, 5 Dec. 2019, moving.tips/bonus-tips-and-tricks/moving-to-a-cold-climate/. Accessed 14 Sept. 2023.

Hesthaven, Mantas. Embarking on an adventure. Unsplash, 11 Sept. 2016, unsplash.com/photos/man-holding-luggage-photo-_g1WdcKcV3w. Accessed 4 July 2024.

Hilfrank, Elizabeth. "Woolly Mammoth." National Geographic Kids, kids.nationalgeographic.com/animals/prehistoric/facts/woolly-mammoth#. Accessed 19 Sept. 2023.

Holmboe-Ottesen, Gerd, and Margareta Wandel. "Changes in Dietary Habits after Migration and Consequences for Health: A Focus on South Asians in Europe." Food & Nutrition Research, vol. 56, no. 1, Jan. 2012, p. 18891, https://doi.org/10.3402/fnr.v56i0.18891. Accessed 13 Sept. 2023.

"Homo sapiens: Media." Britannica, britannica.com/topic/Homo-sapiens/images-videos. Accessed 19 Sept. 2023.

Hough, Emily, and Nathaniel Counts. "How Climate Change Affects Our Mental Health, and What We Can Do About It." The Commonwealth Fund, 29 Mar. 2023, commonwealthfund.org/publications/explainer/2023/mar/how-climate-change-affects-mental-health/. Accessed 13 Sept. 2023.

"How Does Natural Selection Work?" American Museum of Natural Science, amnh.org/exhibitions/darwin/evolution-today/natural-selection-vista#. Accessed 19 Sept. 2023.

"How to have a balanced diet in your new country?" Global Health, 20 Apr. 2020, foyerglobalhealth.com/blog/how-to-have-a-balanced-diet-in-your-new-country/. Accessed 13 Sept. 2023.

"Impact of culture." Rethink Obesity, rethinkobesity.com/disease-progression/impact-of-culture.html#. Accessed 13 Sept. 2023.

National Geographic. "Introduction to Human Migration."

Jordan, Rob. "Is An Extreme Warm West And Cold East Winter The New US Norm?" Innerself, innerself.com/social/environment/weather/13414-is-the-extreme-warm-west-and-cold-east-winter-the-us-norm.html. Accessed 19 Sept. 2023.

"Key Migration Terms." International Organisation for Migration, iom.int/key-migration-terms/. Accessed 7 Dec. 2023.

Kiffel-alcheh, Jamie. "New York." National Geographic Kids, kids.nationalgeographic.com/geography/states/article/new-york. Accessed 20 Sept. 2023.

Kusakabe, Kyoko. "Economic Migrants." Striking Women, striking-women.org/module/types-migration/economic-migrants/. Accessed 9 Sept. 2023.

Lehnardt, Karin. "67 Interesting Facts about Evolution." Fact Retriever, 24 Feb. 2017, factretriever.com/evolution-facts/. Accessed 18 Sept. 2023.

(8) WORK CITED

"Living Outside the Box: New Evidence Shows Going Abroad Linked to Creativity." American Psychological Association, 1 Apr. 2009, apa.org/news/press/releases/2009/04/abroad-creativity/. Accessed 10 Sept. 2023.

"London." History, 7 Mar. 2019, history.com/topics/european-history/london-england/. Accessed 10 Sept. 2023.

Loewe, Laurence. "Genetic Mutation." Scitable, 2008, nature.com/scitable/topicpage/genetic-mutation-1127/#. Accessed 19 Sept. 2023.

MacColl, Andrew. "Directional Selection." Oxford Bibliographies, 19 Nov. 2021, oxfordbibliographies.com/display/document/obo-9780199941728/obo-9780199941728-0049.xml#:~:text=obo%2F9780199941728%2D0049. Accessed 20 Sept. 2023.

Martin, Ryan A., and David W. Pfennig. "Widespread Disruptive Selection in the Wild Is Associated with Intense Resource Competition." BMC Evolutionary Biology, vol. 12, no. 1, 2012, p. 136, https://doi.org/10.1186/1471-2148-12-136. Accessed 5 July 2024.

Massazza, Alessandro. "Explained: How climate change affects mental health." Wellcome, 7 Nov. 2022, wellcome.org/news/explained-how-climate-change-affects-mental-health. Accessed 13 Sept. 2023.

Mathews, Rebecca. "How Did Mammoths Go Extinct?" A-Z Animals, 13 Oct. 2022, a-z-animals.com/blog/how-did-mammoths-go-extinct/#. Accessed 19 Sept. 2023.

McMahon, Mary. "What is Blood Viscosity?" The Health Board, 1 Sept. 2023, thehealthboard.com/what-is-blood-viscosity.htm. Accessed 20 Sept. 2023.

National Geographic Society. "Adaptation." National Geographic, 14 July 2022, education.nationalgeographic.org/resource/adaptation/. Accessed 18 Sept. 2023.

"Natural Selection." American Board, americanboard.org/Subjects/biology/natural-selection/. Accessed 5 July 2024.

Nucleus AI. "[Morning Quote] A Journey of a Thousand Miles Begins With a Single Step." Your Story, 21 Apr. 2023, yourstory.com/2023/04/embrace-journey-one-step-at-time-personal-growth#. Accessed 10 Sept. 2023.

Olivia Dun, François Gemenne, "Defining environmental migration : Why it matters so much, why it is controversial and some practical processes which may help move forward ", REVUE Asylon(s), N°6, november 2008, Exodes écologiques, url de référence: reseau-terra.eu/article847.html

"Potential Net Migration Index." Gallup, news.gallup.com/migration/interactive.aspx. Accessed 20 Sept. 2023.

Relocation, Santa Fe. "The expat diet a shift in food habits." Santa Fe Relocation, 7 Oct. 2019, santaferelo.com/en/moving/news-and-blog/the-expat-diet-a-shift-in-food-habits/# . Accessed 13 Sept. 2023.

Rob. "Hello. Nice to Meet You. Benefits of Meeting New People." Vybe, 19 Apr. 2017, vybe.care/blog/hello-nice-meet-benefits-meeting-new-people/#:~:text=Emotional%20Health,support%20to%20cope%20with%20trauma. Accessed 11 Sept. 2023.

Sawe, Benjamin Elisha. "The Different Types Of Human Migration." World Atlas, 8 Aug. 2018, worldatlas.com/articles/the-different-types-of-human-migration.html. Accessed 9 Sept. 2023.

Sherburne, Morgan. "Homo sapiens: The global 'general specialist.'" University of Michigan, 30 July 2018, news.umich.edu/homo-sapiens-the-global-general-specialist/. Accessed 18 Sept. 2023.

Shiju, Neha. "Wat voor invloed heeft migreren op je gezondheid? Ik schrijf een boek en ik moet een paar anoniem ervaringen toevoegen. Alvast bedankt :))." Quora, Quora, Inc., nl.quora.com/Wat-voor-invloed-heeft-migreren-op-je-gezondheid-Ik-schrijf-een-boek-en-ik-moet-een-paar-anoniem-ervaringen-toevoegen-Alvast-bedankt?. Accessed 20 Sept. 2023.

Simeon, James C., Dona, Giorgia, Voutira, Efthiha, Neuwahl, Nanette, Atak, Idil, Harrell-Bond, Emerieta Barbiera, Ruffer, Galya, Clements, Susan. "Forced Migration." Canadian Association for Refugee and Forced Migration Studies (CARFMS), rfmsot.apps01.yorku.ca/glossary-of-terms/forced-migration/. Accessed 9 Sept. 2023.

Stringer, Chris. "Are Neanderthals the same species as us?" Natural History Museum, nhm.ac.uk/discover/are-neanderthals-same-species-as-us.html. Accessed 19 Sept. 2023.

Tattersall, Ian. "Homo sapiens." Britannica, 27 June 2023, britannica.com/topic/Homo-sapiens/. Accessed 18 Sept. 2023.

WORK CITED

Tea from Culture Tourist. "The Most Beautiful Cities In The Netherlands Besides Amsterdam." Brogan Abroad. 26 Apr. 2023, broganabroad.com/most-beautiful-cities-in-netherlands/. Accessed 12 Sept. 2023.

"The Psychology Of Culture Shock." Verto Education, 30 May 2023, vertoeducation.org/blog/psychology-of-culture-shock/. Accessed 13 Sept. 2023.

Tiwari, Pratishtha. "Alert! Sudden change from hot to cold .." Times of India. 16 May 2019, timesofindia.indiatimes.com/life-style/health-fitness/health-news/alert-sudden-change-from-hot-to-cold-can-be-harmful-to-your-health/articleshow/69354918.cms. Accessed 14 Sept. 2023.

Tousignant, Kaylyn. "Microbial diversity: The key to improving gut health." Microba, 4 Feb. 2020, insight.microba.com/blog/microbial-diversity-the-key-to-improving-gut-health/#. Accessed 13 Sept. 2023.

"Types of movements." Essentials of Migration Management, emm.iom.int/handbooks/global-context-international-migration/types-movements-0#:~:text=At%20one%20extreme%2C%20voluntary%20migration.of%20destination%20for%20admission. Accessed 1 July 2024.

"Understanding Habits and Why They are Important to our Health." Cecelia Health, 9 Dec. 2020, ceceliahealth.com/understanding-habits-and-why-they-are-important-to-our-health/. Accessed 10 Sept. 2023.

"UN Report: Nature's Dangerous Decline 'Unprecedented'; Species Extinction Rates 'Accelerating.'" United Nations, un.org/sustainabledevelopment/blog/2019/05/nature-decline-unprecedented-report/. Accessed 19 Sept. 2023.

"What Is Culture Shock?" International Student Insurance, internationalstudentinsurance.com/explained/mental-health/culture-shock/what-is-culture-shock.php/. Accessed 13 Sept. 2023.

"What is evolution?" Your Genome, yourgenome.org/facts/what-is-evolution/. Accessed 19 Sept. 2023.

"Why Indoor Heaters are Dangerous to your Health and Wallet." NJ Air Pros, njairpros.com/why-indoor-heaters-are-dangerous-to-your-health-and-wallet/. Accessed 14 Sept. 2023.

Young, Julie. "Brain Drain: Definition. Causes. Effects, and Examples." Investopedia, 30 Apr. 2023, investopedia.com/terms/b/brain_drain.asp#:~:text=Brain%20drain%20is%20a%20s/. Accessed 20 Sept. 2023.

INDEX